Geometric
Patterns
from Roman Mosaics
·and how to draw them·

Robert Field

Tarquin Publications

To Joan Ostrolenk who has borne with fortitude the results of my enthusiasm for Roman mosaics.

The start of it all. A design from the courtyard of the museum of Cordoba.

All the photographs and illustrations are by the author with the exception of the photographs on pages 26 and 42 and the drawings on pages 39 and 48 which are reprinted by courtesy of the Sussex Archaeological Society.

© 2017   Robert Field Second Edition

© 1988   First Edition

ISBN Book   978 1 91109 342 8

ISBN EBook 978 1 91109 343 5

Design: Paul Chilvers

Printing: Fuller Davies Ltd. Ipswich

Tarquin

Suite 74, 17 Holywell Hill

St Albans

AL1 1DT, UK

www.tarquingroup.com

Why draw them?

I first became interested in drawing Roman mosaic patterns while on holiday in Southern Spain. I had been to Italica and had photographed many of the mosaics there, realising just what a wealth of interest and different designs there were. From there I went on to Cordoba, where there was a wonderful floor in an open courtyard in the museum. Unfortunately photography was not allowed, and indeed my camera had been taken from me at the entrance. As seems so often to be the case, no postcard or reproduction of that particular design, a feast of running peltae in black, white, red and yellow, was available. I knew that it would be impossible to remember the design and so I analysed it and drew it in my note book, in order to be able to reproduce it on my return home.

Since that holiday I have seen many other mosaics, both at home and abroad and this book is the result of my growing interest in the subject. Most people are attracted first by the figurative panels in mosaics, but my eye has always been taken by the geometric elements of the overall designs. As a result of my experience at Cordoba, together with visits to a number of important sites in Italy, I began to collect and study these geometric patterns and began to realise the intelligence and sophistication of those early artists. Gradually the realisation dawned on me that the geometric patterns were a fascinating study in their own right, not only to see what the Roman artists had achieved, but also as a rich source of design material for the future.

These geometric patterns radiate symmetry and order. Drawing the patterns is not just a question of mechanically copying the work of someone else square by square, but of understanding the underlying structure. The patterns are built up from simple elements which seem to 'grow' and develop in an almost organic or living way.

This book is arranged as a series of drawing exercises. There is no better way of appreciating the skill and imagination of those artists than by drawing their designs yourself. To 'feel' how a cross 'grows' into a swastika pattern which then 'grows' into a complex interlocking design is something which can only be experienced at first hand.

I commend you to try!

Robert Field

It is easy to get started on drawing Roman patterns, because they are largely based on a square grid. It is possible to buy sheets of squared paper in a wide variety of sizes or to draw your own. For each design I have suggested the size of grid required and later in the book, when the designs become more complicated, I have indicated several stages in the work. This helps to bring out the underlying structure.

The finest collection of geometric mosaics in England are at Fishbourne Roman Palace, just outside Chichester and many of the designs are to be found there. At the back of the book is a list of other sites in England where good mosaic patterns may be seen and it is hoped that you will be able to visit some of these.

Let us start with one of the most famous of the mosaics at Fishbourne, the one called 'The Boy on a Dolphin'. However, let us look not at the boy himself, but at the edging. This is perhaps the simplest geometric design of all. A simple chequer board design where the black and white squares alternate.

Usually the chequer board design is not coloured in such a simple fashion and on this page there are two other versions. To draw them you need a grid of 11 x 14 squares for the upper pattern and 10 x 12 squares for the lower.

Both these designs come from a place called Lucus Feroniae, which is just outside Rome. The upper pattern is from a large villa and the lower one from the baths of a house near the Forum.

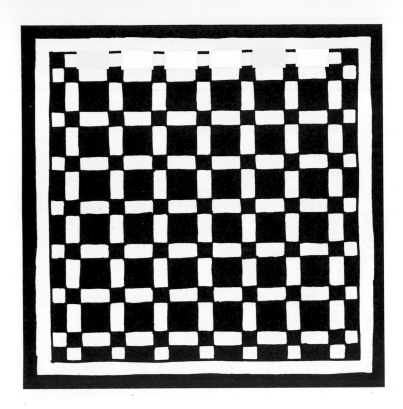

This is another variation on the chequer board idea from Fishbourne, Room N.4.

This sketch shows how to construct it on a 25 x 25 square grid.

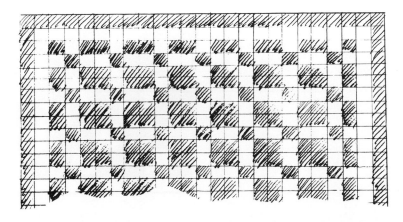

red

grey

This is another variation on the chequer board theme, which is now in the Dorchester Museum. It is in four colours – red, grey, black, and white.

The border of the Orpheus Mosaic at Littlecote in Berkshire is a three colour version of the chequer board design. The colours are black, white and red.

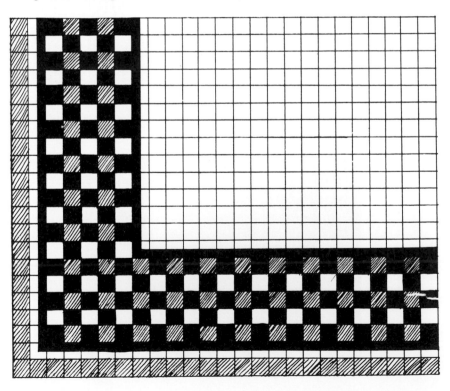

Sometimes different chequer board designs are themselves arranged in a chequer board pattern. Here a design developed from a 5 x 5 square grid is coloured in two alternate ways which are themselves alternated. It is found in the Roman Palace of Fishbourne as a floor of a corridor in the North Wing and as a panel of black and white mosaic in Room W.6 in the West Wing (now covered over).

To construct the pattern above you will need a grid of 25 x 10 squares. Variations (A) and (B) alternate.

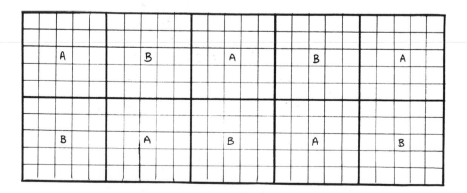

The design above uses the same idea of diagonal squares as the mosaic on the previous page, but alternated with a larger square pattern. It comes from the House of the Mosaic Fountain in Pompeii where it is part of the floor of the atrium surrounding the impluvium.

The centre of this mosaic is a repeated 5 x 5 design, coloured in two opposite ways.

It is a mosaic from Silchester, the Romano-British town of Calleva Atrebatum. It was found in what is thought to be a church. To draw it you will need an 18 x 18 square grid. The diamonds around the edge are alternately black and red. Note that some squares have to be divided in half.

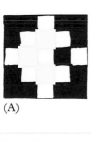

(A)

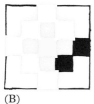

(B)

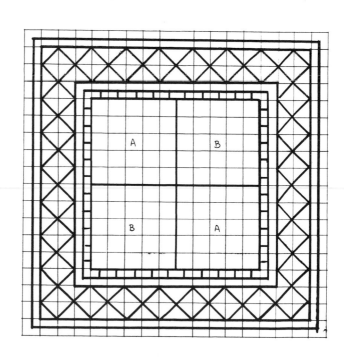

This floor from Silchester, which is now in Reading Museum, is a particularly interesting one. It is an 'all-over meander' pattern with a number of 5 x 5 and 4 x 4 designs within it. It was originally the floor of a house.

Three different 5 x 5 designs are included.

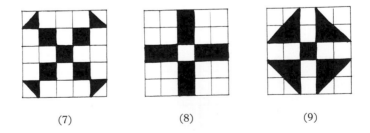

(7)   (8)   (9)

The numbers show how many black squares out of the 25 there are in each of the three designs.

Eight different 4 x 4 designs are included.

Here are a further seven different 4 x 4 designs. They are not included in that floor, but may be seen elsewhere.

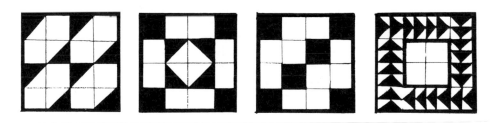

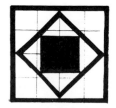
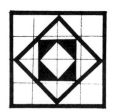

To draw the whole mosaic is rather too complicated, but it is interesting to draw this simplified version. You need a grid of 44 x 32 squares. Start by drawing the outline shown on the page opposite.

Develop the design by adding longer arms to each cross at right angles to make a swastika shape as in (B) below. Add longer arms again at right angles as in (C), until the pattern takes the form shown below.

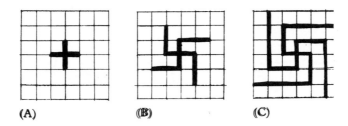

(A) (B) (C)

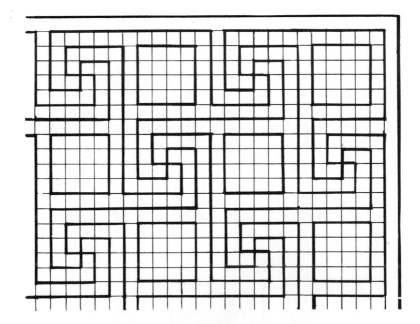

Since this is a 'modified' pattern you can now add any of the 4 x 4 designs on the previous page to fill the squares.

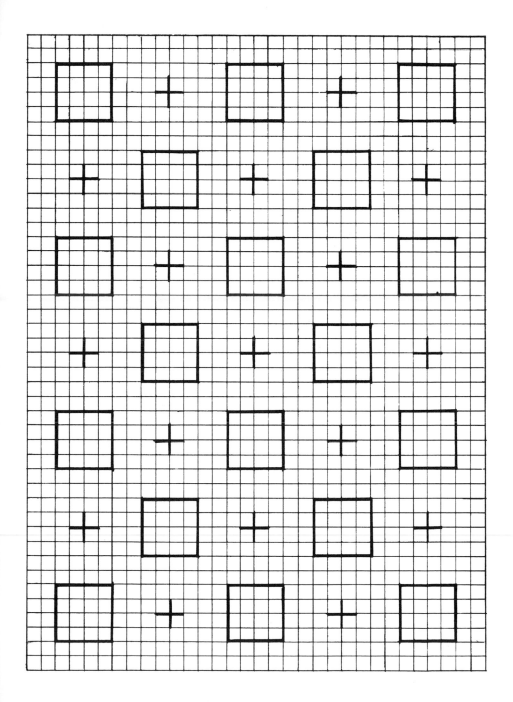

On the Silchester mosaic on page 12 there is also a single 3 x 3 design.

Without the diamond in the middle, it is frequently used, often being repeated hundreds of times. The design below is of the floor of one of the cubiculae (bedrooms) of the Hospitalium (Guest Room) of Hadrian's Villa at Tivoli outside Rome.

This same simple 3 x 3 design is also used as an edging in some of the Fishbourne mosaics and as a design for the complete floor of one room (N.19)

To draw the design of the floor of Room N.19 at Fishbourne you will need a grid of 11 x 11 squares. Follow the stages shown below.

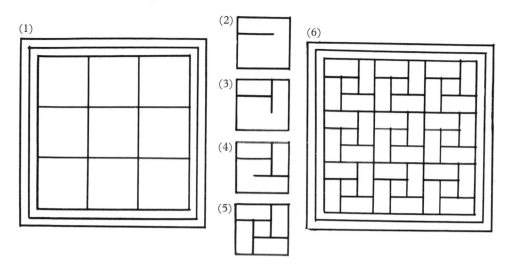

(1) (2) (3) (4) (5) (6)

Here are two more 3 x 3 designs, this time coloured in two opposite ways.

They are used alternately on this mosaic from the Church of Saint Sebastian in Rome. To draw it, use a 12 x 12 square.

There is a meander mosaic similar to the one on pages 14 & 15 in Herculaneum. It is on the floor of the Forum Baths. It has a smaller square within the larger ones with a design in each of them. Most of these represent objects and are not geometric like the one from Silchester.

The Corridor mosaic at Bignor Roman Villa in Sussex also uses this type of meander design.

To draw the central portion of the Bignor corridor mosaic you will need a grid which is 45 squares in width. The pattern repeats itself every 24 squares so you can make it as long as you want. The squares and the pattern inside them are red.

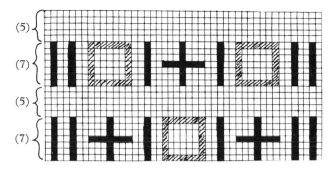

(5)
(7)
(5)
(7)

(A)   First draw up the lines, crosses and squares (red).

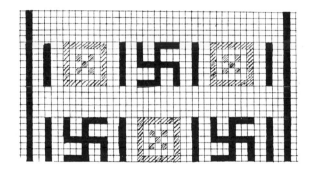

(B)   Fill in the pattern inside the squares and make the crosses into swastikas. Join the outside lines and extend them.

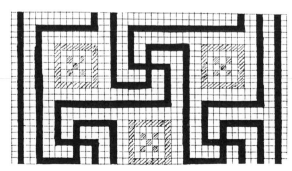

(C)   Extend the lines and join them. You can now repeat the pattern as many times as you want to. If you want to add the guilloche border, see page 63 and add a further 12 squares on each side.

The meander pattern is often used as a border. In its simplest form is usually called a 'fret' or 'Greek key' pattern.

Here are two such patterns. To draw them you will need grids of squares in the form of strips.

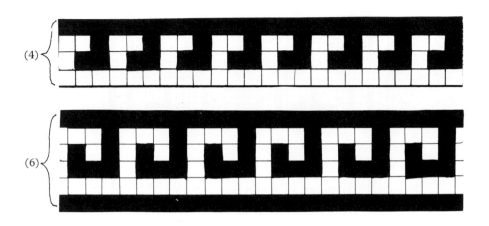

These next two examples are 'castellated patterns', which are found at Bignor.

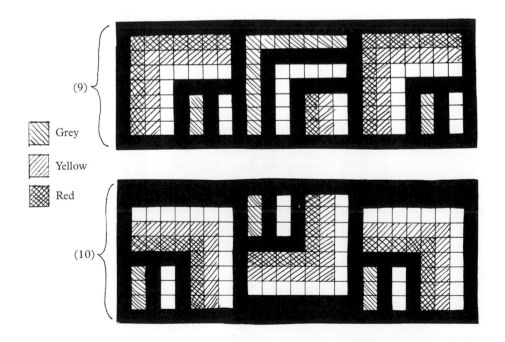

Grey

Yellow

Red

On the next three pages there are examples of meander border patterns. To draw this one you will need a grid which is 13 squares wide. Follow the five stages indicated below.

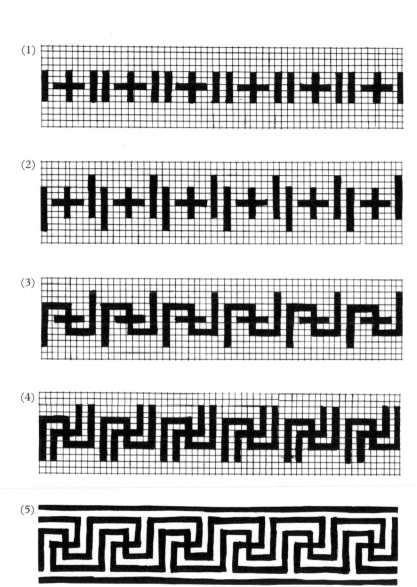

This one also needs a strip which is 13 squares wide.

(1)

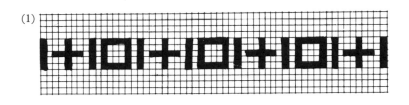

(2)

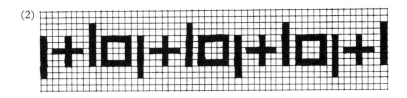

(3)

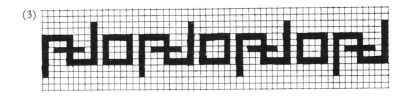

(5)

(4)

This is a more complicated pattern and needs a strip which is 17 squares wide.

(1)

(2)

(3)

(4)

(5)

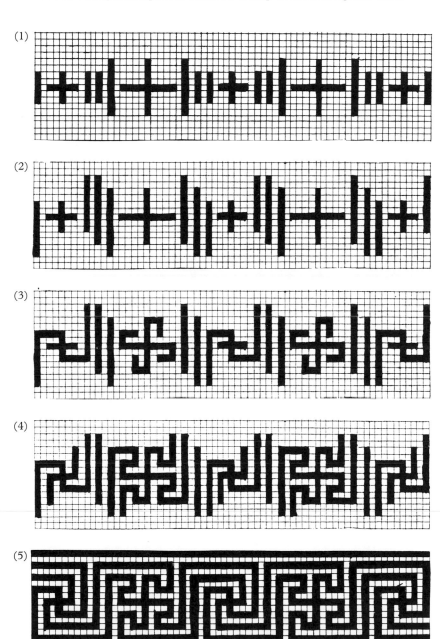

To draw the border to the 'Sea God' mosaic you will need a 60 x 60 square grid.

(1)

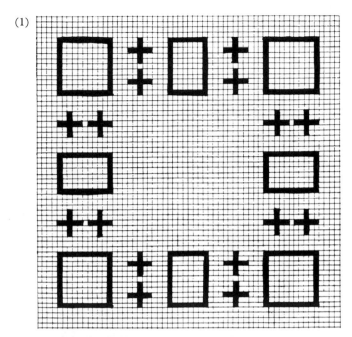

(2)

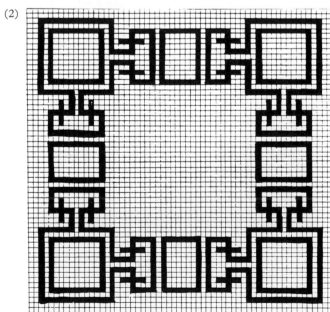

This is the famous 'Sea God' mosaic from St Albans, which is now in the Verulamium Museum. It is called the Sea God Mosaic after the figure in the middle. It was found in House IV in the Romano-British city. It has a meander pattern around the squares and rectangles enclosing the decorative motifs.

(3)

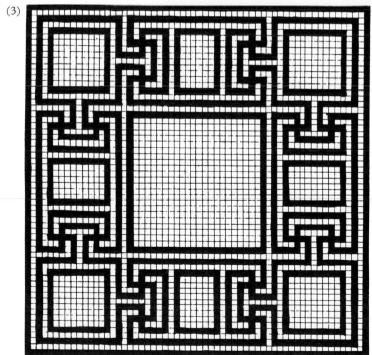

At Fishbourne Roman Palace in Room W.3. (now covered over), there is a complicated meander floor pattern. You can see in the photograph that some of the squares in the design also contain a further meander pattern.

To draw it you will need a grid of 75 x 47 squares. Take great care to draw your squares and crosses in the correct places. It is a most satisfying one to complete.

(1)

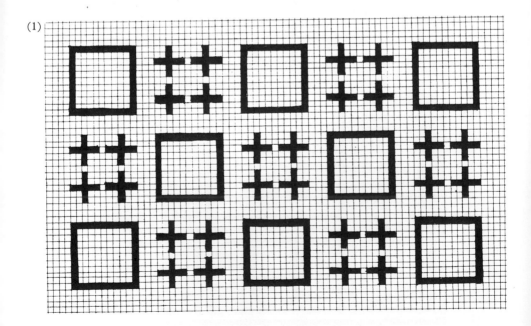

(2)

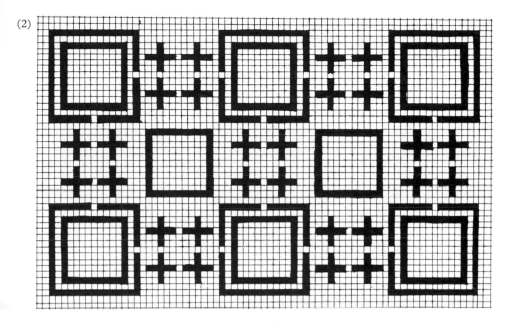

(3)

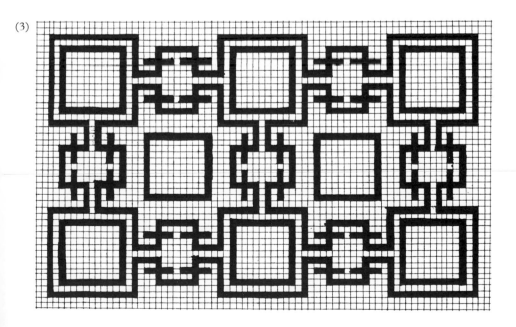

(4)

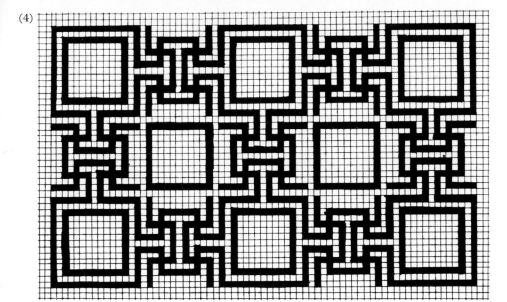

(5)

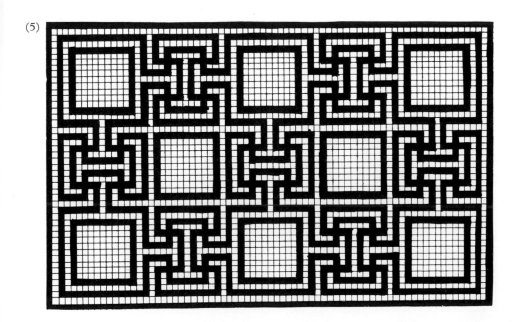

Most of the Fishbourne mosaics are black and white. The white part forming the main section of the floor with the black lines outlining the design. In Room N.21 however, the pattern is predominantly black with the outlines in white. The squares in the middle are red and grey.

To draw it, follow the instructions below and on the next two pages. You will need a grid which is 51 x 117 squares.

(A)
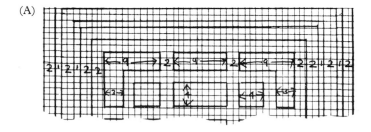

Lay out the whole
floor.

(C)

Colour it as indicated.

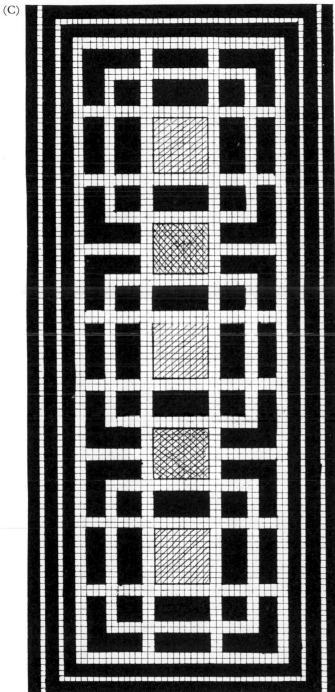

 red

grey

Many designs use diagonal lines to divide the squares.

If you put two lines of these together other interesting patterns begin to emerge.

The pattern above is used at Fishbourne as a 'doormat' in the outside border of the 'Boy on a Dolphin' mosaic.

The pattern below comes from a mosaic in the Forum of the Corporations in Ostia which was the ancient port of Rome.

When both diagonals are drawn in a square, then this opens up many more design possibilities.

The first example below is from a house in the Forum of Lucus Feroniae.

This second example is from the Hospitalium of Hadrian's Villa at Tivoli.

This border pattern of triangles is a common one. It is used at Fishbourne in Room W.6.

It can be formed either from a single square or from a group of four squares.

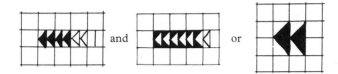

It is often used in conjunction with meander patterns.

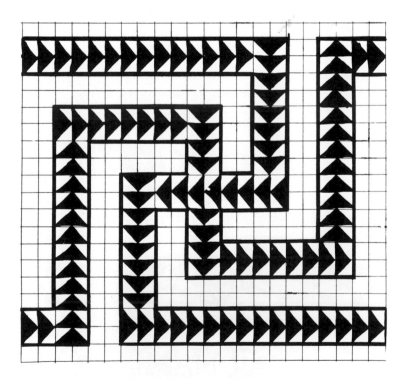

Many simple patterns described earlier in this book can be enlarged and decorated in this way to create original designs of your own.

This mosaic which makes great use of the triangle pattern is now in the British Museum. It comes from Saint-Romain-en-Gal in France. The centre panel shows Silvanus and his dog and the roundels show Bacchus, Silenus, Pan and a satyr. To draw it you will need a 27 x 27 grid.

This mosaic from Pompeii, which is now in the Naples Museum also makes much use of triangles. They are very cleverly obtained from a square. To draw it, you need a grid of 14 x 14 squares.

Once you have drawn the square as in (1), cut it out as in (2) and then colour as in (3).

(1)
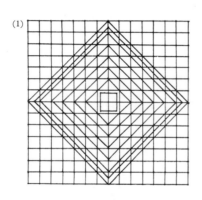

(2)
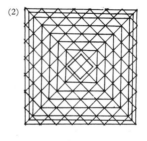

(3)

This first century pattern from Room N.3 at Fishboume is a nice one to draw. You need a grid of 18 x 30 squares. Follow the four stages indicated below.

(1)

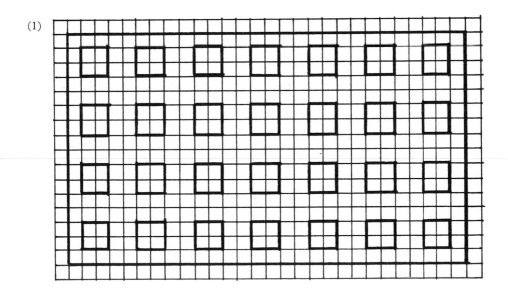

(2)

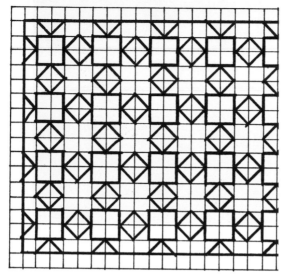

(3)

(4)

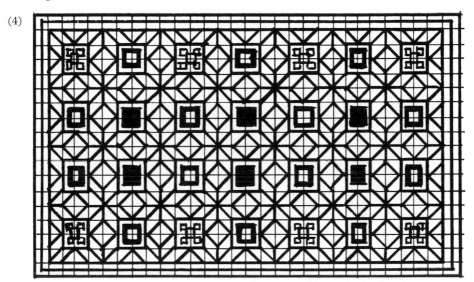

Also at Fishbourne in Room N .12 is one of the best preserved mosaic patterns. Below is a drawing of the complete floor and there is a coloured picture of part of it on page 42 .

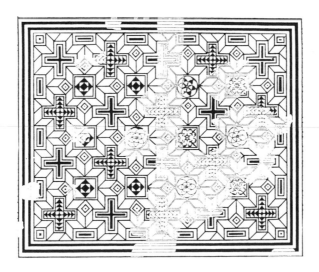

To draw the mosaic in Room N.12, you need a grid of 68 x 83 squares. Firstly draw the outline design given below and then follow the numbered stages opposite.

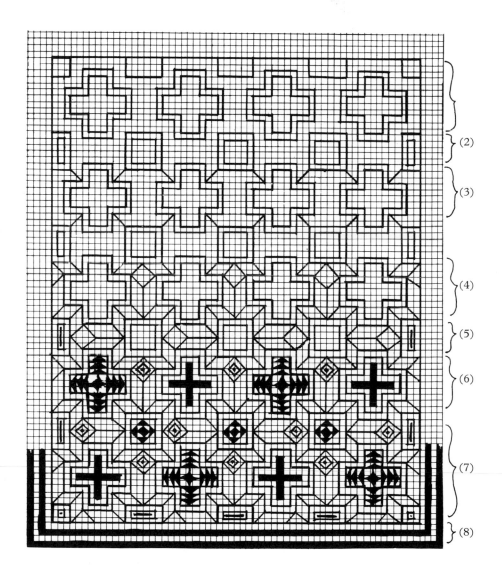

(2)

(3)

(4)

(5)

(6)

(7)

(8)

Room N.12 at Fishbourne.

This mosaic from Room W.8. at Fishbourne is now covered over. To draw it you will need a grid of 26 x 38 squares. Follow the numbered stages carefully as it is a complex design.

(A) Draw out the whole design.

(B) Develop the detail in the order shown by the numbers 1 to 6

(C)This is the completed design.

The two mosaics below come from the Villa of the Volussi in Lucus Feroniae. You need to work out the size of the grid that you will need for each of them before you start.

In Room 6 at Bignor Roman Villa there is a geometric mosaic. It is formed from two squares surrounded by lozenge or diamond shapes, which give the impression of three-dimensional cube shapes. Between the two major designs based on the squares there is a rectangle with a vase in its centre. The designs of the two squares can each be drawn up on a grid of 6 x 6 squares.

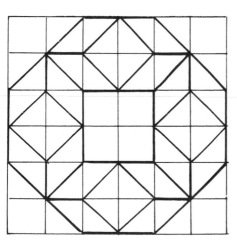

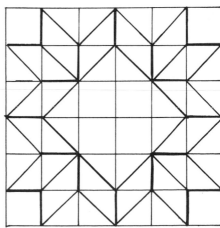

In 1979 the 'Boy on a Dolphin' mosaic at Fishbourne Roman Palace was lifted for conservation work as it was begining to deteriorate. Underneath was found part of a black and white mosaic. It has been lifted and relaid at Fishbourne. It consists of a panel of 16 squares, each with a different geometric design, surrounded by a 'Fortress' mosaic. This consists of a representation of city walls with gates in the four sides and towers on the corners. The arches around the gates contain some red and grey pieces.

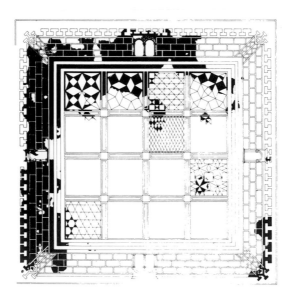

Four of the designs within the squares are drawn here.

(A)

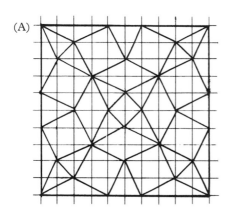

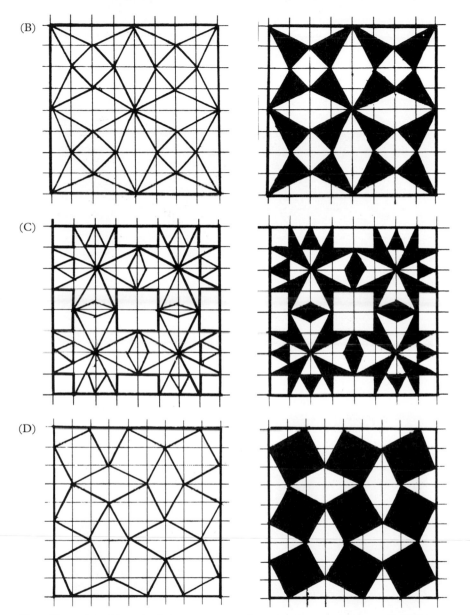

You will see that although all the squares are the same size, one has been divided into a 10 x 10 square grid, one into a 9 x 9 and the other two into 8 x 8. To draw this particular mosaic you will need to start with a plain piece of paper and then divide the edges of the squares into 8, 9 or 10 divisions as needed.

Since the contents of some of the squares are missing, you can invent some of your own designs to fill the gaps.

There are many patterns which give an illusion of three dimensions. A favourite Roman floor design is to make a cube-like shape from three diamonds which form a hexagon. This floor which is cut from different coloured marbles is in the Temple of Apollo in Pompeii.

The floor below is from the Villa of the Volussi in Lucus Feroniae. It is a black and white mosaic and gives a similar three-dimensional illusion.

To draw the three-dimensional cube design you will need a grid of 27 x 30 squares.

Here are two further developments which give the illusion of three-dimensions. The first is from Turkey and the second from Syria. Both have been drawn on an 18 x 18 square grid.

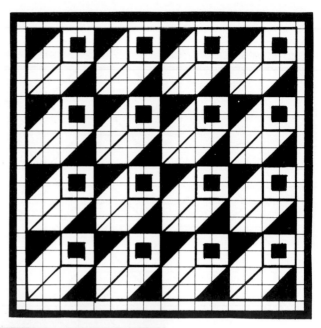

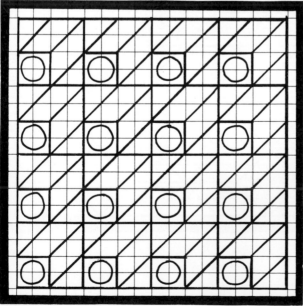

This border pattern also gives the illusion of three dimensions. You will need a grid which is 19 squares wide. Each repeat is 24 squares long.

Experiment with different colours to heighten the three-dimensional effect.

These two designs use a hexagonal rather than a square grid. You can use isometric paper or, if you can find it, one with a hexagonal grid.

The first pattern is from the House of Livia in the Forum in Rome and the second is from Italica in Southern Spain.

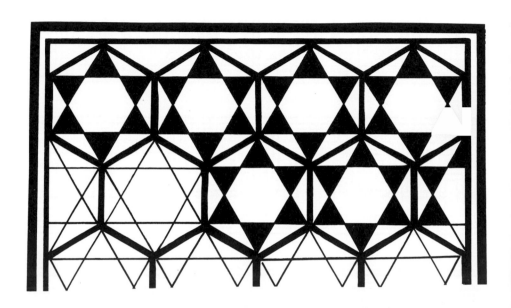

So far we have been looking at straight line shapes. We will now look at some designs that use circles and curved lines. For this design of overlapping circles you will need some isometric paper. It occurs with the white background in Hadrian's Villa at Tivoli and with a black background in the Baths of Caracalla in Rome. Use isometric paper and compasses and follow the sketch below.

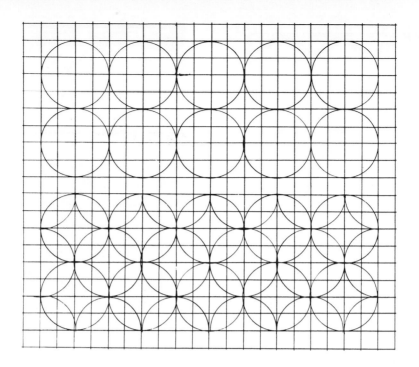

Here is another way of overlapping or inter-linking circles. It is used on a variety of mosaics to be found in Britain. The one below is from the Orpheus Mosaic at Littlecote.

Here are three more mosaics that use this method of over-lapping circles.

This one is from Rockbourne Villa in Hampshire.

This floor is in the Baths suite at Chedworth Villa in Gloucestershire.

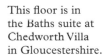

This design is now in the British Museum.

Here are two more patterns using circles. The sketches show how to draw them.

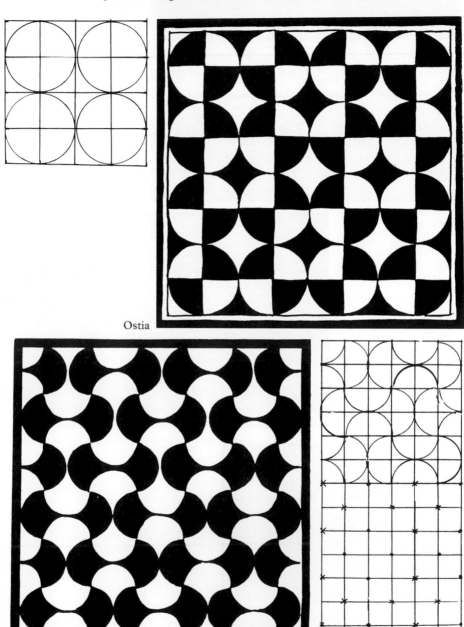

Ostia

Ostia

Here are two more designs that use curved lines. They have to be drawn freehand on a square grid background.

Ostia

Baths of Caracalla, Rome.

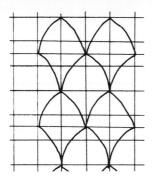

This scale or spade shape has to be drawn on a rectangular grid. A simple way to draw such a grid is to use 1cm squared paper and then to divide every other row into half. Then draw the curved lines freehand.

Both these designs come from Ostia but are used very often by the makers of mosaics throughout the Roman Empire.

Even more popular as a design element than the spade shape is the pelta. It is the shape derived from the shape of the shields said to have been used by the Amazons. It is also sometimes described as an axe shape.

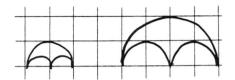

It is combined in many different ways, some of which are shown here. One of the most popular is the combination of four to make a 'swastika pelta'.

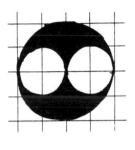 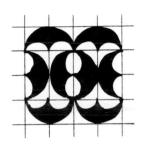

  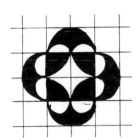

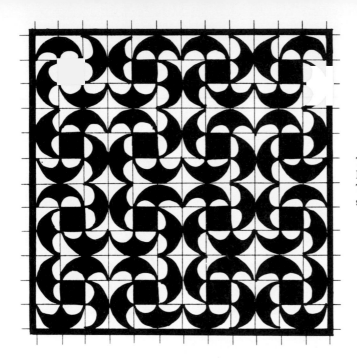

This pelta design is from Hadrian's Villa at Tivoli. It uses a 14 x 14 square grid.

This design from a house in Ostia also uses a 14 x 14 square grid.

These border patterns are probably the most common of all mosaic designs and can be seen on sites all over the Roman Empire. They have to be drawn freehand on a square grid. Each design is a repetition of the shape shown above it. These three designs are known as 'guilloche' border patterns.

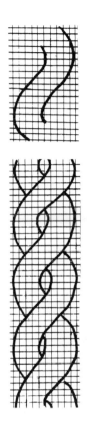

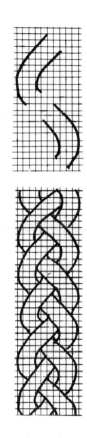

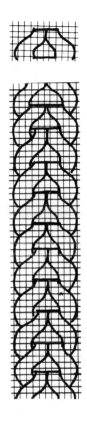

This knot shape is also a very common design. It is known by various names – guilloche knot, duplex knot, Solomon's knot, endless knot, lovers' knot and single knot.

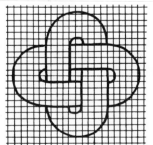

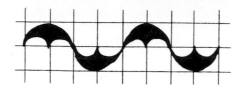

And finally, let us end with the design from the museum at Cordoba which started it all. It is a design called a 'Running Pelta'.